SPARKS OF LIFE

Chemical Elements that Make Life Possible

# IRON and the TRACE ELEMENTS

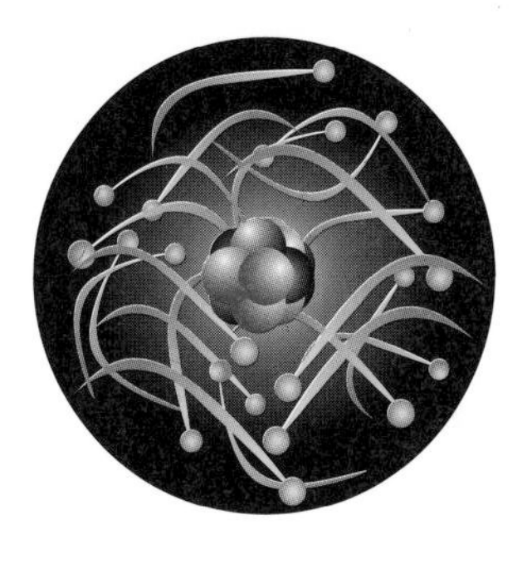

by

Jean F. Blashfield

A Harcourt Company

Austin New York

www.raintreesteckvaughn.com

Special thanks to our technical consultant,
Philip T. Johns, Ph.D.
University of Wisconsin—Whitewater, Wisconsin



**Development: Books Two, Inc., Delavan, Wisconsin**
Graphics: Krueger Graphics, Janesville, Wisconsin
Interior Design: Peg Esposito
Photo Research: Margie Benson
Indexing: Winston E. Black

**Raintree Steck-Vaughn Publisher's Staff:**
Publishing Director: Walter Kossmann    Project Editor:
Design Manager: Richard A. Johnson



**Library of Congress Cataloging-in-Publication Data:**

Blashfield, Jean F.
Iron and the Trace Elements / by Jean F. Blashfield.
p. cm. — (Sparks of life)
Includes bibliographical references and index.
Summary: Defines trace elements and discusses how they affect life on the planet.
ISBN 0-7398-4359-1
1. Iron--Juvenile literature. 2. Iron--Physiological aspects--Juvenile literature. 3. Trace elements--Juvenile literature. 4. Trace elements--Physiological aspects--Juvenile literature. [1. Trace elements.] I. Title.
QD181.F4 B54 2001
546'.621--dc21

2001031676

Printed and bound in the United States.
1 2 3 4 5 6 7 8 9 LB 05 04 03 02 01

PHOTO CREDITS: B.I.F.C. cover; ©Richard Carlton/Visuals Unlimited 10; ©D. Cavagnaro/Visuals Unlimited 22; ©John D. Cunningham/Visuals Unlimited 48; ©Rowell Galen/CORBIS 32; ©Stanley L. Gibbs/Peter Arnold, Inc. 37; ©Holt Studios/Nigel Cattlin 20; ©Richard P. Jacobs/JLM Visuals 17, 21, 26, cover; ©Bert Krages/Visuals Unlimited 16; ©Leonard Lessin/Peter Arnold cover; NOAA 29; ©Charles O'Read/CORBIS 25; ©David M. Phillips/Visuals Unlimited 18; ©Harry J. Przekop, Stock Shop/Medichrome cover; ©Roger Ressmeyer/CORBIS 46; ©James W. Richardson/Visuals Unlimited 50; Science VU-LBL/Visuals Unlimited 40; ©Clyde H. Smith/Stock Shop/Medicrome cover; ©Sinclair Stammers/Science Photo Library cover; ©Larry Stepanowicz/Visuals Unlmited 38; ©Rich Treptow/Visuals Unlimited 8; U.S. Department of Energy/NREL 52; USDA Agricultural Research Service 12, 13, 14, 24, 28, 31, 33, 44, 52, 54(left); VU/SIU/Visuals Unlimited 39; ©Charles D. Winters/PhotoResearchers 43, 54(right)

# CONTENTS

Fe

# Periodic Table of the Elements

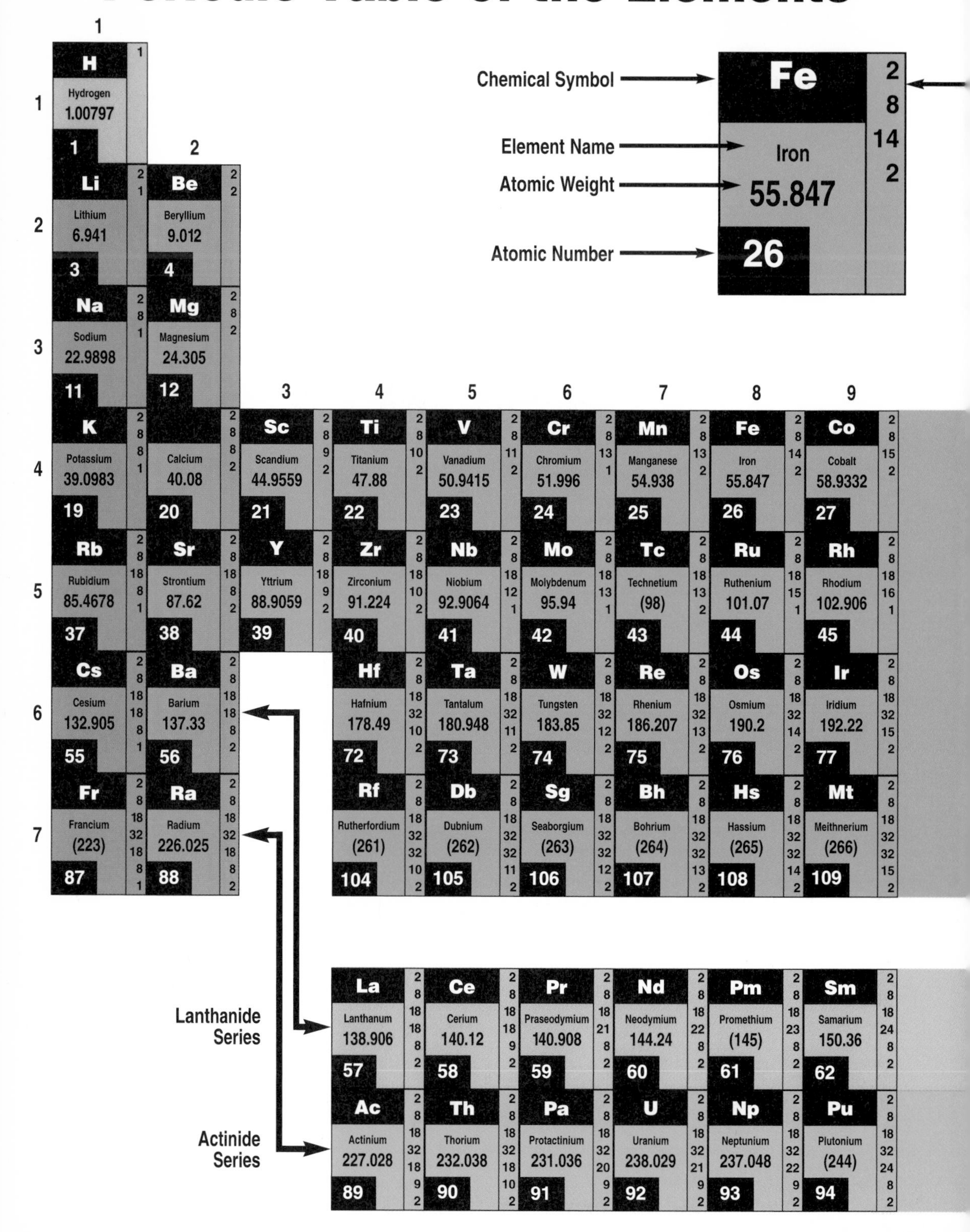

Number of electrons in each shell, beginning with the K shell, top.

See next page for explanations.

| 10 | 11 | 12 | 13 | 14 | 15 | 16 | 17 | 18 |
|---|---|---|---|---|---|---|---|---|
| | | | | | | | | **He**<br>Helium<br>4.0026<br>2<br>2 |
| | | | **B**<br>Boron<br>10.81<br>5<br>2 3 | **C**<br>Carbon<br>12.011<br>6<br>2 4 | **N**<br>Nitrogen<br>14.0067<br>7<br>2 5 | **O**<br>Oxygen<br>15.9994<br>8<br>2 6 | **F**<br>Fluorine<br>18.9984<br>9<br>2 7 | **Ne**<br>Neon<br>20.179<br>10<br>2 8 |
| | | | **Al**<br>Aluminum<br>26.9815<br>13<br>2 8 3 | **Si**<br>Silicon<br>28.0855<br>14<br>2 8 4 | **P**<br>Phosphorus<br>30.9738<br>15<br>2 8 5 | **S**<br>Sulfur<br>32.06<br>16<br>2 8 6 | **Cl**<br>Chlorine<br>35.453<br>17<br>2 8 7 | **Ar**<br>Argon<br>39.948<br>18<br>2 8 8 |
| **Ni**<br>Nickel<br>58.69<br>28<br>2 8 16 2 | **Cu**<br>Copper<br>63.546<br>29<br>2 8 18 1 | **Zn**<br>Zinc<br>65.39<br>30<br>2 8 18 2 | **Ga**<br>Gallium<br>69.72<br>31<br>2 8 18 3 | **Ge**<br>Germanium<br>72.59<br>32<br>2 8 18 4 | **As**<br>Arsenic<br>74.9216<br>33<br>2 8 18 5 | **Se**<br>Selenium<br>78.96<br>34<br>2 8 18 6 | **Br**<br>Bromine<br>79.904<br>35<br>2 8 18 7 | **Kr**<br>Krypton<br>83.80<br>36<br>2 8 18 8 |
| **Pd**<br>Palladium<br>106.42<br>46<br>2 8 18 18 | **Ag**<br>Silver<br>107.868<br>47<br>2 8 18 18 1 | **Cd**<br>Cadmium<br>112.41<br>48<br>2 8 18 18 2 | **In**<br>Indium<br>114.82<br>49<br>2 8 18 18 3 | **Sn**<br>Tin<br>118.71<br>50<br>2 8 18 18 4 | **Sb**<br>Antimony<br>121.75<br>51<br>2 8 18 18 5 | **Te**<br>Tellurium<br>127.6<br>52<br>2 8 18 18 6 | **I**<br>Iodine<br>126.905<br>53<br>2 8 18 18 7 | **Xe**<br>Xenon<br>131.29<br>54<br>2 8 18 18 8 |
| **Pt**<br>Platinum<br>195.08<br>78<br>2 8 18 32 17 1 | **Au**<br>Gold<br>196.967<br>79<br>2 8 18 32 18 1 | **Hg**<br>Mercury<br>200.59<br>80<br>2 8 18 32 18 2 | **Tl**<br>Thallium<br>204.383<br>81<br>2 8 18 32 18 3 | **Pb**<br>Lead<br>207.2<br>82<br>2 8 18 32 18 4 | **Bi**<br>Bismuth<br>208.98<br>83<br>2 8 18 32 18 5 | **Po**<br>Polonium<br>(209)<br>84<br>2 8 18 32 18 6 | **At**<br>Astatine<br>(210)<br>85<br>2 8 18 32 18 7 | **Rn**<br>Radon<br>(222)<br>86<br>2 8 18 32 18 8 |
| **(Uun)**<br>(Ununnilium)<br>(269)<br>110<br>2 8 18 32 32 17 1 | **(Unu)**<br>(Unununium)<br>(272)<br>111<br>2 8 18 32 32 18 1 | **(Uub)**<br>(Ununbium)<br>(277)<br>112<br>2 8 18 32 32 18 2 | | | | | | |

**COLOR KEYS**

| | | | | |
|---|---|---|---|---|
| Alkali Metals | Transition Metals | Nonmetals | Metalloids | Lanthanide Series |
| Alkaline Earth Metals | Other Metals | Noble Gases | Actinide Series | |

| | | | | | | | | |
|---|---|---|---|---|---|---|---|---|
| **Eu**<br>Europium<br>151.96<br>63<br>2 8 18 25 8 2 | **Gd**<br>Gadolinium<br>157.25<br>64<br>2 8 18 25 9 2 | **Tb**<br>Terbium<br>158.925<br>65<br>2 8 18 27 8 2 | **Dy**<br>Dysprosium<br>162.50<br>66<br>2 8 18 28 8 2 | **Ho**<br>Holmium<br>164.93<br>67<br>2 8 18 29 8 2 | **Er**<br>Erbium<br>167.26<br>68<br>2 8 18 30 8 2 | **Tm**<br>Thulium<br>168.934<br>69<br>2 8 18 31 8 2 | **Yb**<br>Ytterbium<br>173.04<br>70<br>2 8 18 32 8 2 | **Lu**<br>Lutetium<br>174.967<br>71<br>2 8 18 32 9 2 |
| **Am**<br>Americium<br>(243)<br>95<br>2 8 18 32 25 8 2 | **Cm**<br>Curium<br>(247)<br>96<br>2 8 18 32 25 9 2 | **Bk**<br>Berkelium<br>(247)<br>97<br>2 8 18 32 26 9 2 | **Cf**<br>Californium<br>(251)<br>98<br>2 8 18 32 28 8 2 | **Es**<br>Einsteinium<br>(254)<br>99<br>2 8 18 32 29 8 2 | **Fm**<br>Fermium<br>(257)<br>100<br>2 8 18 32 30 8 2 | **Md**<br>Mendelevium<br>(258)<br>101<br>2 8 18 32 31 8 2 | **No**<br>Nobelium<br>(259)<br>102<br>2 8 18 32 32 8 2 | **Lr**<br>Lawrencium<br>(260)<br>103<br>2 8 18 32 32 9 2 |

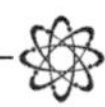

# A Guide to the Periodic Table

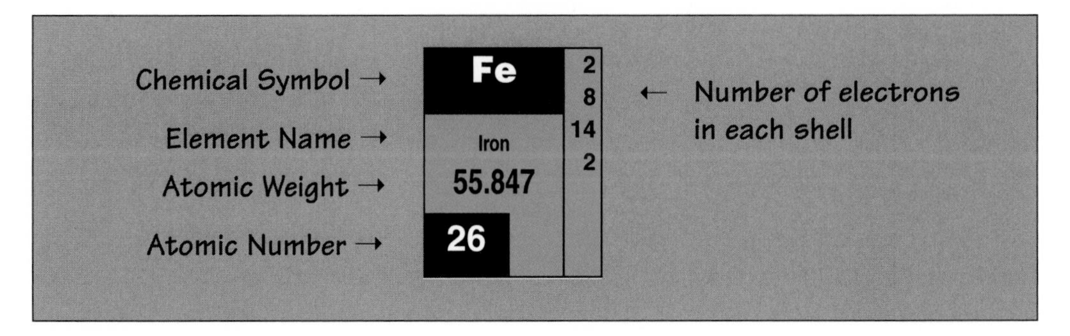


**Symbol** = an abbreviation of an element name, agreed on by members of the International Union of Pure and Applied Chemistry. The idea to use symbols was started by a Swedish chemist, Jöns Jakob Berzelius, about 1814. Note that the elements with numbers 110, 111, and 112, which were "discovered" in 1996, have not yet been given official names.

**Atomic number** = the number of protons (particles with a positive electrical charge) in the nucleus of an atom of an element; also equal to the number of electrons (particles with a negative electrical charge) found in the shells, or rings, of an atom that does not have an electrical charge.

**Atomic weight** = the weight of an element compared to carbon. When the Periodic Table was first developed, hydrogen was used as the standard. It was given an atomic weight of 1, but that created some difficulties, and in 1962, the standard was changed to carbon-12, which is the most common form of the element carbon, with an atomic weight of 12.

The Periodic Table on pages 4 and 5 shows the atomic weight of carbon as 12.011 because an atomic weight is an average of the weights, or masses, of all the different naturally occurring forms of an atom. Each form, called an isotope, has a different number of neutrons (uncharged particles) in the nucleus. Most elements have several isotopes, but chemists assume that any two samples of an element are made up of the same mixture of isotopes and thus have the same mass, or weight.

**Electron shells =** regions surrounding the nucleus of an atom in which the electrons move. Historically, electron shells have been described as orbits similar to a planet's orbit. But actually they are whole areas of a specific energy level, in which certain electrons vibrate and move around. The shell closest to the nucleus, the K shell, can contain only 2 electrons. The K shell has the lowest energy level, and it is very hard to break its electrons away. The second shell, L, can contain only 8 electrons. Others may contain up to 32 electrons. The outer shell, in which chemical reactions occur, is called the valence shell.

**Periods =** horizontal rows of elements in the Periodic Table. A period contains all the elements with the same number of orbital shells of electrons. Note that the actinide and lanthanide (or rare earth) elements shown in rows below the main table really belong within the table, but it is not regarded as practical to print such a wide table as would be required.

**Groups =** vertical columns of elements in the Periodic Table; also called families. A group contains all elements that naturally have the same number of electrons in the outermost shell or orbital of the atom. Elements in a group tend to behave in similar ways.

Group 1 = alkali metals: very reactive and so never found in nature in their pure form. Bright, soft metals, they have one valence electron and, like all metals, conduct both electricity and heat.

Group 2 = alkaline earth metals: also very reactive and thus don't occur pure in nature. Harder and denser than alkali metals, they have two valence electrons that easily combine with other chemicals.

Groups 3–12 = transition metals: the great mass of metals, with a variable number of electrons; can exist in pure form.

Groups 13–17 = transition metals, metalloids, and nonmetals. Metalloids possess some characteristics of metals and some of nonmetals. Unlike metals and metalloids, nonmetals do not conduct electricity.

Group 18 = noble, or rare, gases: in general, these nonmetallic gaseous elements do not react with other elements because their valence shells are full.

# MINERALS FOR HEALTH

Most living things are made up of just eleven fundamental ingredients, or chemical elements. Amazingly, though, only four of these elements make up 99 percent of the bodies of complex living organisms such as human beings. These four are carbon, nitrogen, oxygen, and hydrogen. Oxygen and hydrogen atoms combine to make water, which forms the sheer bulk of all living bodies.

This is a chemist's view of 9 of the 18 elements recognized as essential to human life.

At least 18 other chemical elements—often called minerals by nutritionists—make up the remaining 1 percent of living tissue. Ninety percent of that 1 percent is formed by only seven elements. These seven elements are called macronutrients (*macro* means "large") in that we need to obtain relatively large quantities of them —usually through food—in order to replenish the supplies needed by our bodies. These seven are sodium, potassium, calcium, magnesium, phosphorus, sulfur, and chlorine.

## One-Thousandth of the Body

That does not leave much—just one-tenth of 1 percent. Even so, the eleven elements that make up that tiny amount are vital for life and health. Often called micronutrients (*micro* means "tiny"), they are the trace elements. Alphabetically, the micronutrients for humans are chromium, cobalt, copper, fluorine, iodine, iron, manganese, molybdenum, selenium, and zinc.

That list included only ten elements. The eleventh is boron. Boron is a macronutrient for plants but has not been determined to be even an essential micronutrient for mammals. Plants need fewer micronutrients than animals do—just seven. They are iron, manganese, zinc, boron, molybdenum, copper, and chlorine. Chlorine is a micronutrient for plants but a macronutrient for mammals. (Chlorine has a book of its own in this series.)

One definition of a trace element is that less than 100 milligrams per day of the element is required for good health. This is in contrast to the macronutrients, which are needed in comparatively large amounts every day. To be a micronutrient, an element must account for less than 0.05 percent of the human body. The most abundant trace element in human beings is iron, which makes up about 0.006 percent of the body's weight.

## The Metals

Like iron, most micronutrients are metals that are important to industry. Manganese, chromium, copper, molybdenum, and cobalt are among the metals

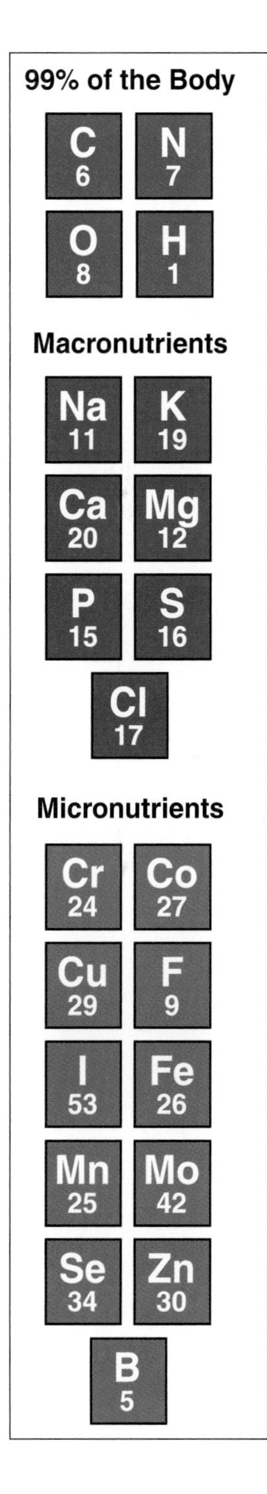

added to iron to make various kinds of steel. Several of these metals are naturally magnetic—iron, nickel, and cobalt.

The Sacagawea dollar coin consists entirely of trace metals. Its inner core is copper. Surrounding this is an outer golden-colored coating that is an alloy—a mixture of metals—of 77 percent copper, 12 percent zinc, 7 percent manganese, and 4 percent nickel.

Combinations of metallic trace elements often serve even more important purposes. For example, various alloys of cobalt, chromium, and (sometimes) molybdenum are used to make new hip and knee joints to replace the natural joints in people whose joints have been destroyed by disease or injury.

## What Do They Do?

It is estimated that 94 percent of all fatal diseases in humans involve trace nutrients in some way. This does not necessarily mean that a deficiency of one or another actually causes the disease but that it plays a role in the formation or development of the disease.

Even so, the body has an amazing ability to take care of itself by recognizing when it needs a nutrient. In such cases, the intestine will absorb more of the needed element from food being digested than it normally does. The body can also take some nutrients out of "storage." Most excess nutrients are stored in the liver. However, there is a limit to the amount of the element that can be stored.

**Parsley contains several essential nutrients, especially copper, magnesium, iron, and boron.**

Some of the foods that contain the most trace elements are called "protective foods." They contain the elements that play a role in the immune system of the body. The immune system is all the cells, organs, and bodily

processes that help fight off invasion by bacteria, viruses, and other outside agents. Iron, zinc, selenium, and copper are found in protective foods.

Most trace minerals are necessary for the functioning of enzymes in the body. Enzymes are catalysts—substances that are involved in speeding up a chemical change but are themselves not changed by the process. Enzymes are the "control levers" of the processes of the body. Hundreds of enzymes have been identified. They all carry out very specific tasks. Scientists usually add the ending "–ase" to a chemical name to indicate that it is an enzyme. Many enzymes include metallic trace elements.

A trace mineral that works with a specific enzyme is called a cofactor of the enzyme. About one-third of known enzymes require a cofactor. These proteins may also require cofactors to be activated, or set into action.

Most nutrients, including trace minerals, work in living things as ions. Ions are atoms or molecules that are missing one or more electron or have taken on one or more additional electrons. In an atom, the protons in the nucleus, which have a positive electric charge, balance the electrons in orbit around the nucleus; the electrons are negative. In an ion, though, this electrical balance no longer exists. The ion carries either a positive charge, because it has more protons than electrons, or a negative charge because it has more electrons than protons.

For example, an atom of zinc has only two electrons in its outer, or valence, shell. Such an atom is not stable, but it would be stable without those two electrons. Then the new outer shell would be complete. It would have 18 electrons, all it can hold. A zinc atom, then, readily gives up its two valence electrons, resulting in a positive charge. It is written $Zn^{2+}$.

**A zinc atom becomes an ion, $Zn^{2+}$, by losing two electrons.**

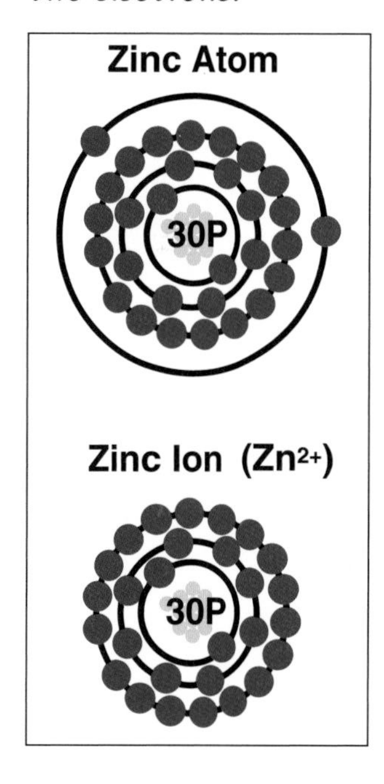

**Scientists study the substances in food to determine which ones we need for health.**

## What Do We Need?

The National Academy of Sciences (NAS) makes recommendations about how much of a particular nutrient humans need to be healthy. For essential minerals, the NAS recommends a specific amount for daily intake. This is called the DRI, for "Daily Recommended Intake," or the RDA, for "Recommended Daily Allowance."

Often, micronutrients are safe and necessary at tiny amounts but can be poisonous, or toxic, to the body at higher amounts. Chromium is one of the elements that can become toxic. For such elements, the NAS has declared a UL, or tolerable upper intake level. Taking in more than the UL can be dangerous to health. For this reason, people who take extra doses of a mineral supplement may actually be doing themselves great harm.

The body usually adjusts to taking in too much of a mineral by excreting it in the urine. But a point is reached at which the kidneys—which purify body fluids before excreting them as urine—cannot handle the excess mineral. The kidneys could begin to shut down, which would be fatal if the excess dosage were continued.

Iron, copper, and zinc have a very narrow range at which they are helpful to the body and not harmful. These three minerals interact. If you take in too much copper, the body stops

using iron and zinc properly. Too much iron can prevent zinc from being absorbed.

## Missing from Our Food

The more our food is processed—treated with chemicals or handled by machines to change it from its original state—the more trace elements disappear. You have probably heard the phrase "junk foods." The term means that although these snack foods taste good, they have little real nutritional value beyond calories. They are mostly sugar and fat.

Some of the things we eat regularly are not really junk foods, but they do not contain all the nutrients they should. Whole grains are called "whole" for a good reason—they contain all the nutrients that occur naturally in them. After wheat is milled, the resulting white flour is missing an estimated 40 percent of its chromium, 86 percent of its manganese, 76 percent of its iron, 89 percent of its cobalt, 68 percent of its copper, 78 percent of its zinc, and 48 percent of its molybdenum.

**The whole grains from wheat contain many nutrients that are removed when wheat is milled into white flour.**

Obviously, whole grains are considerably better for you than milled white flour. "Enriched" white flour means that several vitamins, plus iron, have been added to the flour. But other trace elements are still missing.

Whole milk contains many micronutrients. When the butterfat is removed from it to make skim milk, some of those micronutrients disappear too. Selenium, molybdenum, and manganese are removed, along with half of the chromium.

Refined white sugar is what's left over after

virtually all the nutrients have been removed. The nutrients are left in the dark-brown gooey syrup called molasses that is extracted. Molasses is rich in chromium, manganese, copper, zinc, and molybdenum, and contains some selenium. White sugar contains almost no minerals. People today eat much less molasses than they used to—and more white sugar.

## A New Revolution

The scientific advances that have led to the development of more bountiful crops have been called the "green revolution." The green revolution has fed many people of the developing world. However, as scientists developed seeds that grew so plentifully, many of the trace elements needed for complete nutrition disappeared from the resulting crops.

Children are usually the first to suffer and die from malnutrition. Even when they get enough food in quantity, they still may not get sufficient trace elements to fight off disease or to grow properly. Each year, more than 6 million children under the age of five die because of malnutrition. Around the world, the two micronutrients most likely to be lacking in children's diet are iron and iodine.

Many of the same scientists who were involved in the green revolution are now trying to start a nutrient revolution. They hope to develop new hybrid crops that will return the micronutrients to the foods grown in large quantities that are available to people in developing countries.

**Agricultural scientists hope to add nutrients to crops that are planted around the world.**

# IRON: THE BIG ONE

The Iron Age is the name given by archaeologists to the period of early history when human beings first learned to work with the metal. At that time, pure iron was quite scarce and especially precious because all iron on Earth is bound up with other elements. The iron used by these ancient humans was pure metal that came to Earth in meteorites.

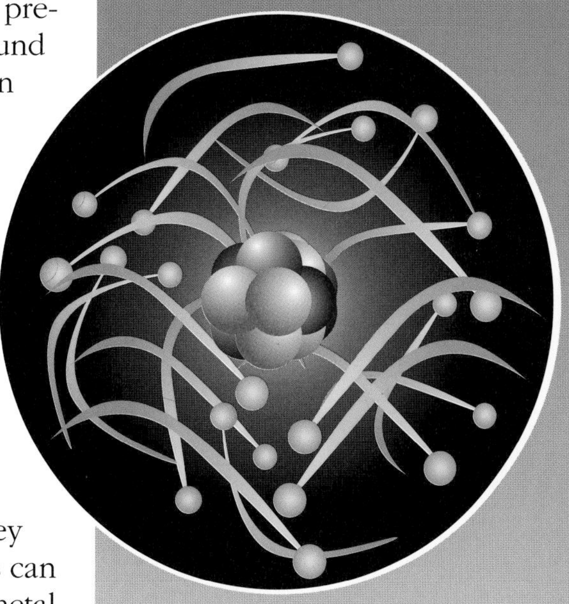

By 4,000 years ago, humans had learned to heat iron ore to obtain the metal. The Industrial Revolution started 300 years ago when people finally learned to obtain iron from iron ore cheaply and in large quantities. Soon afterward, they found that by adding carbon (C, element #6) to iron, they could produce steel. Different metals can be alloyed with steel to produce metal with different qualities. Chromium and nickel, for example, make stainless steel.

## Iron the Element

Iron has atomic number 26. Its symbol, Fe, comes from its old Latin name, *ferrum*. Iron belongs to Group 8 (also called VIII) in the Periodic Table of the Elements. Elements in this group have two electrons in the outer, or valence, shell. Iron reacts easily with other elements, giving up those two electrons in order to make a stable outer electron shell. An element that gives up electrons is said to be *oxidized* (even though the process may not involve oxygen).

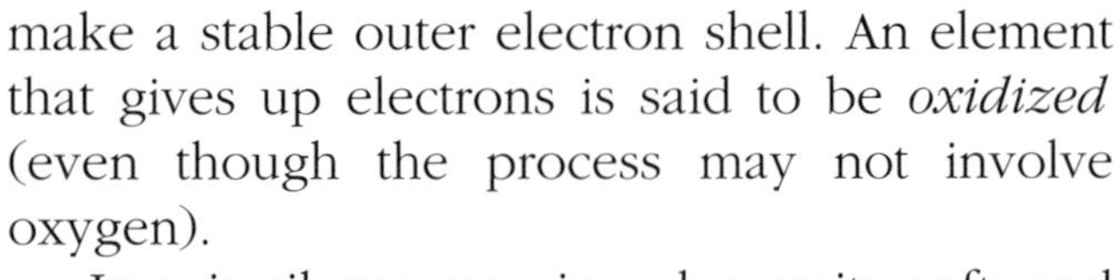

Iron is silvery gray in color, quite soft, and easy to work, even at room temperature. Pure iron reacts with the oxygen in air or water to become iron oxide, $Fe_2O_3$, which is rust. Rusting is the most common example of oxidation. Rust is an orange-colored powdery coating that can form on iron and then crumble away, taking iron with it.

**Metallic iron rusts in the damp soil. This iron tank was dug up from the ground because it was leaking.**

At 5.6 percent of Earth's crust, iron is the fourth most abundant element after oxygen, silicon (Si, element #14), and aluminum (Al, #13). Iron is the second most abundant metallic element, after aluminum. It is present in all soils and many rocks combined with oxygen (as oxides) and silicon (as silicates).

Iron is a fairly unusual element in that it can be made magnetic. Electrons have a tiny electric charge. In a single atom, the electrons generally spin in different directions, and these charges cancel each other out. However, if a piece of iron is put within a magnetic field, the charges line up in the same direction and get stronger. It will then attract other pieces of iron.

**Compasses respond to the magnetic pull of a chunk of naturally magnetic iron rock, magnetite.**

A naturally magnetic rock called magnetite, or lodestone, is an iron oxide, $Fe_3O_4$. Another, hematite, is $Fe_2O_3$. They are both important sources of metallic iron. During Earth's long history, iron may have melted and cooled several times. As it cooled, the crystals of iron retained a position aligned with Earth's magnetic field. Geologists use this fact to determine changes in the planet during its history.

## Iron in Our Bodies

Iron's effect on humans has been studied probably more than that of any other element. This research started when an 18th-century researcher accidentally found that he could move dried blood with a magnet. He realized that there must be iron in blood. As early as 1832, doctors began to give tablets containing microscopic bits of iron to patients with "pale" blood.

The strong red color of blood comes from a substance called heme. About 65 percent of all iron in the body goes into making heme. It is the central compound in hemoglobin, the substance that transports oxygen from the lungs to every cell. About 90 percent of a red blood cell is hemoglobin.

The heme molecule is similar to the chlorophyll molecule in plants. They both start with a metallic atom—iron for animals

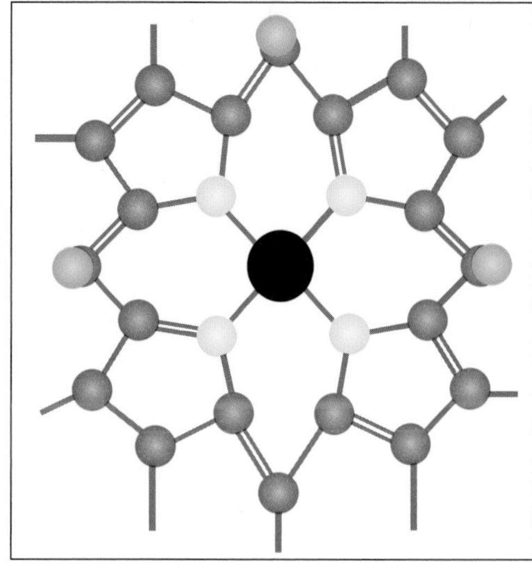

Red blood cells (above) are the main carriers of oxygen in the body. Oxygen is carried by the compound called hemoglobin. The central part of the hemoglobin molecule (right) is built around a single iron atom. It is surrounded by nitrogen (yellow), carbon (red), and hydrogen (green).

and magnesium for plants. This atom is surrounded by a ring structure of nitrogen and carbon called a porphyrin ring. Other atoms are attached in side chains.

A specialized kind of hemoglobin is called myoglobin. It carries oxygen to muscles, preparing them for use. The iron-centered heme molecule in myoglobin is what makes muscle tissue red. It is easily visible in red meat such as beefsteak.

The body needs a continuous supply of iron because red blood cells gradually age and die and have to be replaced. The best natural sources of iron are the organs of livestock, such as liver and kidneys. Fortunately for vegetarians, iron is also found in eggs, spinach, and legumes (dried beans). The iron in food is in the form of $Fe^{3+}$ ions. During digestion, the ions take up one electron and become $Fe^{2+}$ ions, which are absorbed into the bloodstream. In the blood, $Fe^{2+}$ ions turn into $Fe^{3+}$ ions again.

$Fe^{3+}$ ions are used in an important protein called transferrin. Transferrin carries iron to the liver, spleen, and bone marrow. In those places, iron is removed from transferrin and incorporated into hemoglobin. Once iron is absorbed by the body, it is not excreted. Loss of iron generally occurs only from bleeding.

## The Main Micronutrient

There are two kinds of iron in the food we eat. Heme iron is combined with carbon, making what is called organic iron, which comes primarily from animal foods—meat, fish, and eggs. Oysters are especially good at supplying heme iron. The other kind of iron is nonheme iron, which is found primarily in plants, though also in animal products. Kelp, or seaweed, has the highest iron content.

It is estimated that a person in a modern industrialized society takes in about 15 to 20 milligrams of iron every day. Less than 1 mg of that iron is actually absorbed by a man's body, while about 1.5 mg is absorbed by a woman's body. A woman absorbs more because iron is lost during menstruation.

Iron is absorbed more readily by a person who has the right amount of hydrochloric acid in the stomach. For that reason, a person who frequently takes antacids for heartburn might end up with iron deficiency.

Iron deficiency is the most common nutritional disease in the world. Most iron quickly combines with other ions and is lost to use by the body. Iron ions also combine with substances in food. All of these combinations make the iron insoluble, meaning that it cannot form useful ions.

Anemia is a term used for a variety of conditions in which a person has either too few red blood cells or not enough hemoglobin. First described 3,500 years ago, anemia has many causes and many consequences. Iron-deficiency anemia in children can result in poor growth and delayed sexual maturing. In adults, it shows up first as fatigue.

In developing countries, iron deficiency occurs most often when people rely solely on whole grains and legumes for

**This strawberry plant shows that its soil is deficient in iron by the lightness of the green color on some leaves.**

nutrition. In richer, industrialized countries such as the United States, the person most likely to develop anemia is a teenage girl. She needs iron both for growth and because of loss of blood during menstruation.

The recommended intake of iron each day for men is 10 milligrams. It is 15 milligrams for women during the years that they menstruate. After menopause, the requirement drops to 10.

Iron is an essential micronutrient for plants as well as animals. It is necessary for the production of chlorophyll, and it is a constituent of enzymes. Iron deficiency in plants shows up as loss of color between the veins of the leaves.

Some processed foods have iron added to them to improve their nutritional value. Grains processed into cereal, for example, often have microscopic bits of actual metallic iron incorporated into the flakes. They are dissolved by stomach acid, which turns them into ions that can be taken into the blood.

Livestock also need iron, especially baby pigs. Piglets grow so quickly that they need a great deal of iron, yet their mother's milk gives them only about 15 percent of their daily needs. Also, piglets raised on the concrete floor of a barn need iron more than piglets that are free to root around in the soil.

# ZINC AND COPPER: INTERTWINED METALS

Zinc, element #30 with symbol Zn, and copper, element #29 with symbol Cu (from the Latin *cuprum*), lie next to each other in the Periodic Table of the Elements. The two are often mixed together, or alloyed, to take advantage of their characteristics. In living things, the functions of copper and zinc are often intertwined. After iron, these two trace elements are needed in the largest quantities.

## Metallic Zinc

Zinc is a fairly soft, bluish-white metal. It has been known since ancient times in the oldest known alloys, brass and bronze. Brass is zinc and copper. Bronze is zinc and tin (Sn, for the Latin *stannum,* element #50).

A new metallic copper coating is being applied to this dome. It will gradually change to green copper oxide, as shown here on the lower dome.

**This man is wearing zinc oxide in a sunblock ointment to prevent skin cancer.**

Pure zinc was first known in India 1,000 years ago, when metalworkers managed to separate it from its natural alloy, bronze. It is found in the ore called zinc blende, which is sulfide sphalerite.

Zinc serves as the outer casing of some batteries. The casing is the negative electrode, the material that gives up electrons to create a flow of electrons, which is electricity. The zinc ions, $Zn^{2+}$, then react with oxygen, forming zinc oxide, $ZnO_2$.

The zinc oxide formed in a battery has no use there. But zinc oxide produced otherwise is the most important zinc compound. As an ointment, it has long been used to treat acne. Because it blocks the sun's rays from reaching the skin, it is often worn as a sunblock. Zinc oxide is also used in the vulcanizing of rubber to make it harder and more difficult to melt.

Because stainless steel is expensive to manufacture, steel may be coated with zinc to protect it from corrosion. Coating steel with zinc is called galvanizing. In the past, pots and pans were galvanized to keep them from rusting. Foods that stood in such pots for a long time absorbed zinc from the metal.

## Zinc in the Body

Because it affects the making of proteins (major compounds of the body), zinc is a vital element. It helps the body fight off infections and repair wounds. Without enough zinc, the body cannot use phosphorus properly. Phosphorus is needed by the cells to burn sugar and produce energy.

Zinc also appears to play a vital role in intelligence and memory. The U.S. Agricultural Research Service conducted a study in China in which schoolchildren with too little zinc in

their bodies were given supplemental zinc. The children's scores in tests of memory, perception, and reasoning improved.

Red blood cells, which require both iron and copper, also contain zinc, in the enzyme carbonic anhydrase. Iron helps the blood pick up oxygen and carry it throughout the body. Zinc, however, helps the blood pick up carbon dioxide, $CO_2$, the waste product of energy production in the cells. It helps carry $CO_2$ to the lungs, where it is exhaled.

Most trace elements are stored in the liver, but zinc is concentrated in bones. It is also found in skin and hair.

Zinc exists in at least 70 metallic enzymes found throughout the body. It plays a role in the production of DNA (deoxyribonucleic acid), the material of genetic inheritance. Without adequate zinc, a man's body does not properly produce sperm, the cells that join with a woman's eggs to produce children.

## Intake of Zinc

Zinc comes primarily from high-protein foods, especially beef. Vegetarians can meet their daily requirement by making sure they consume milk and eggs along with plenty of legumes. Peanuts and popcorn also contain zinc.

Daily intake of zinc from a normal diet is probably between 7 and 16 milligrams. The lower end of this range is below the recommended daily requirement, however. Women need about 12 milligrams a day, and men need about 15 milligrams.

Many food supplements contain zinc, so that if a person takes several different kinds, he or she may inadvertently get too much zinc. A chronic overdose may affect the cholesterol level in the blood. It can decrease the amount of HDL (or "good") cholesterol and increase the amount of LDL (or "bad") cholesterol. This combination can lead to heart problems.

Because zinc has become a popular cold prevention remedy, some people are taking extra zinc. They need to be sure to

**Zinc is among the heavy metals used by industry that have damaged the environment in many places. However, scientists are studying certain plants for their ability to grow in such places and to absorb zinc from the soil.**

get plenty of copper as well, to offset the tendency of zinc to prevent copper from being absorbed.

## Zinc in Plants

Zinc was found to be an essential element for plant growth as early as 1869. It was more difficult to prove that it was essential for humans because the element is found in virtually all foods. Such research had to wait until foods low in zinc could be developed in the late 1950s.

As the ion $Zn^{2+}$, zinc is an essential micronutrient in plants. It is vital in the early growth of plants and in their development toward reproduction. Grains, in particular, do not develop sufficient seeds without it. Zinc is also important in the plant hormone called auxin, which is responsible for increasing length in stems and size in leaves. Though zinc is found in soil, deficiency can lead to several diseases that affect nut and fruit trees.

As with most elements, though, too much zinc can be bad. Runoff in water from zinc mines and smelters can seriously harm plant and aquatic life. As a result, animals that live on or near the water containing zinc can also be harmed.

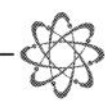

## Copper

Copper, silver (Ag, for *argentum,* element #47), and gold (Au, for *aurum,* #79) were the first known metals. They are the only metals that exist as naturally occurring pure metal in the ground. The three metals occur in the same group, or vertical column, of the Periodic Table—Group 11 (also known as IB), all of which have one electron in the valence orbit.

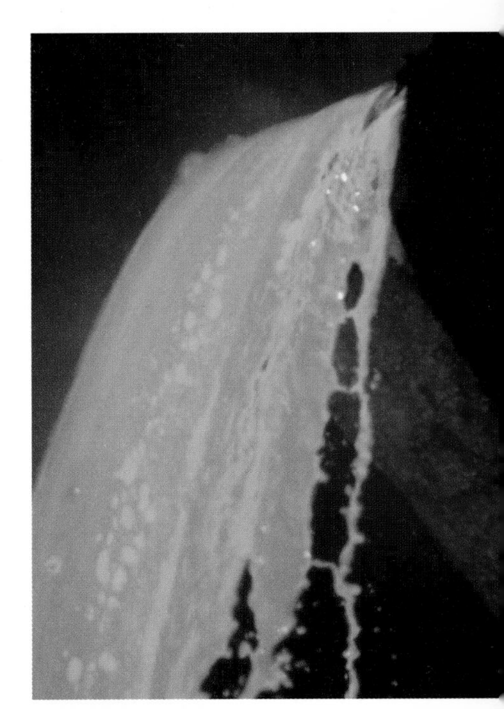

**Metallic copper being processed at a Chilean copper mine**

Pure, or "native," copper was made into tools as early as 9000 B.C. Copper was also probably the first metal extracted from its ore by the process called smelting. In these first instances of smelting, the blue-green ore called malachite, which is copper carbonate, $Cu_2CO_3(OH)_2$, was heated. The heat separated the water and the carbon in the ore, leaving metallic copper behind.

By about 2,000 years ago, copper was produced from copper sulfate, $CuSO_4$, by running water containing the mineral across pieces of iron. The release of the metallic copper from the copper sulfate was thought to be magic. Clearly, the iron had changed into copper. It would be 1,600 years before the chemical reaction involved was fully understood. The two electrons in the valence shell of the iron had moved to the copper:

$$Cu^{2+} + Fe^{0} \rightarrow Cu^{0} + Fe^{2+}$$

The largest copper deposits in the world are found in Chile, where they came to the surface in volcanoes. When copper combines with oxygen and other elements, it turns beautiful shades of green and blue.

**Malachite is one of the main ores of copper.**

Metallic copper is ductile, which means that the metal can be worked into a fine wire without breaking. This trait led to copper's most important use—as wire to carry electricity. The copper used for this purpose has a tiny amount of oxygen deliberately left in it because its presence seems to improve the conduction of electricity.

Copper also conducts heat well. Its melting point is quite high, at least 1,000°C (1,932°F). As a result, copper makes good cooking utensils.

Besides iron, copper is used in more alloys than any other metal. In addition to bronze, one of the oldest alloys is brass, which is mainly copper and zinc. The more copper there is in the brass, the easier it is to work. Copper and nickel form a variety of useful alloys, including one called nickel silver, which contains no silver whatsoever.

Copper lasts longer than iron because it does not combine as easily with oxygen, so it is less likely to rust and wear away. For this reason, buildings often have an exposed layer of copper. The copper turns greenish in color when an oxide layer, called a patina, forms. Unlike rust, this layer remains intact, protecting the copper underneath it.

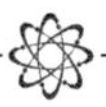

## Copper in Living Things

As we learned in the last chapter, iron deficiency can result in anemia. Anemia can also be caused by a copper deficiency, since both copper and iron are critical ingredients in the formation of hemoglobin. Copper is also involved in the production of collagen, the group of proteins that make up skin, cartilage, and tendons, which connect muscle to bone.

Copper in the body is most heavily concentrated in the brain and the liver. However, the greatest amount of copper in the body—about 25 percent of it—is stored in skeletal muscle. It is required for the making of normal bone tissue.

As is true of most trace elements, copper plays an important role in enzymes. Among the metallic enzymes containing copper are those that assist in blood clotting, the formation of melanin, which makes color in skin, and the use of vitamin C.

One of the most vital enzymes is called cytochrome oxidase. It contains both copper and iron. Without it, the use of oxygen for respiration could not take place.

Certain enzymes in the body are called antioxidants. They prevent other substances from oxidizing, or reacting to the free oxygen that is released during metabolism. Heart muscle, for example, is aged by oxidation. Several enzymes protect heart muscle from oxidation, and at least two of these include copper.

As humans get older, many do not receive enough copper. This may contribute to the aging process. Copper is an important part of the protein ceruloplasmin, which is built as the body is attacked by various oxidants. It has been described as a "fire extinguisher" for the body.

## The Body's Need

For many decades, few people in Western countries suffered from copper deficiency because public water supplies traveled

through copper pipes and picked up some of the metal. In recent years, though, water pipes have been made of plastics rather than copper. This means that people now need to pay more attention to copper in their diets.

It is recommended that adults receive between 1.5 and 3 milligrams of copper each day. Like iron, copper is found in important quantities in oysters. It is also found in nuts (especially Brazil nuts), cocoa powder, and organ meats, such as liver and kidneys. The actual dosage of copper a person should get, however, depends on how much zinc is taken in. Normally, one should get about 10 times as much zinc as copper.

Copper and molybdenum also interact in the bodies of mammals. For example, an animal with a copper deficiency may absorb harmful amounts of molybdenum. The two elements must be in balance. Copper deficiency is a common problem for cattle all over the world. For this reason, copper is often added to livestock feed.

Plants require copper in the form of the ion $Cu^{2+}$ for the production of chlorophyll and seeds. Copper also plays a critical role in the manufacture of enzymes in plants. Copper deficiency reveals itself when newer leaves turn yellow and die.

Oysters (at center bottom) and Brazil nuts (at bottom right) are excellent sources of copper.

# THE MIDDLE WEIGHTS: MANGANESE, BORON, FLUORINE

Three elements are needed in small but not tiny amounts. Their daily requirements are still measured in milligrams rather than micrograms. One of these elements, manganese, is a metal. Boron is a metalloid, meaning that it has some of the characteristics of a metal and some of a nonmetallic element. Fluorine is a nonmetal.

These nodules, or lumps, of pure manganese metal were found on the floor of the ocean.

## Manganese

Manganese was discovered and isolated in 1774 by Johan Gottlieb Gahn, a Swedish geologist. However, Carl Wilhelm Scheele, the discoverer of oxygen, had announced even earlier that he thought manganese was an element. The element received its name from the Latin *magnes,* meaning "magnet." Gahn powdered some manganese dioxide, $MnO_2$, which was then known as pyrolusite. When he heated it in a closed container, the metal separated out as silvery-white bits.

Manganese oxide is the tenth most common mineral compound in Earth's crust. Manganese itself makes up about 0.09 percent of the crust. Most manganese is taken from its ore, pyrolusite, which is black in color. Some scientists are hoping that it might someday become practical to harvest the lumps, or nodules, of pure manganese found on the ocean floor.

Most manganese, which is very similar to iron, is used in making steel. There are several reasons why this is so. Manganese cleans sulfur out of molten iron and makes iron easier to shape when it is hot. It also prevents iron from shattering when it is subjected to a great impact. However, like iron itself, manganese rusts when it combines with the moisture in air.

Scientists who develop computers are working on using manganese molecules to construct hard drives. They have found that manganese allows them to store 30 trillion bits of information in a space of only 1 square centimeter.

## The Manganese Ion

Two varieties of manganese are present in the soil—Mn(II) and Mn(IV). These varieties are similar to the types of iron described earlier. Mn(II), more often known as the ion $Mn^{2+}$, is the form in which manganese is taken into a plant. It plays a role in the formation of carbohydrates. Plants without enough manganese develop dead streaks on the leaves. Excess manganese in soil can be harmful to crops. Some soils in Hawaii contain more manganese than soils found anywhere else on the planet.

In humans, manganese is an antioxidant that plays a role in just about all the major activities of the body—the production of

Scientists examine the roots of eastern gamagrass, one of the main prairie grasses. The specialized roots, which hold water through long droughts, may make manganese and other metals harmless to other plants.

energy, digestion, reproduction, and the transmission of nerve impulses for normal brain function. In childhood, the element is critical to bone growth. Young children who do not receive enough manganese can develop short bones and thick joints.

In the 1930s, manganese was found to be essential to humans, though it was already known to be necessary to livestock, especially pigs and chickens. The human body needs between 2 and 5 milligrams of manganese daily. Avocados, blueberries, seaweed, and whole grains are particularly good sources. It's also present in large quantities in tea, so that people in tea-drinking countries, such as England, get more manganese than Americans do.

As with many trace elements, too much manganese can be toxic. Workers at manganese mines sometimes take in too much from the air they breathe. Excess manganese can prevent the body from processing all the iron it needs. In extreme cases, excess manganese can lead to hallucinations and outbreaks of violent behavior. Symptoms of the condition called manganism are similar to those of the nervous condition called Parkinson's disease. The body's reaction to manganese is being studied in the hope that it might lead to a cure for Parkinson's.

Boron is being dropped from an airplane to help put out a forest fire.

## Boron

Boron, symbol B, is atomic number 5. In the Periodic Table, boron is at the head of Group 13, also called IIIA. This position means that it has three electrons in its valence shell. Boron does not easily give away those three electrons or take on more from another element.

In 1808, several chemists worked on boron separately. The French chemists Joseph-Louis Gay-Lussac (who pioneered balloon flight during his study of gases) and Louis-Jacques Thenard isolated boron as an element. They did so by reacting boric oxide, $B_2O_3$, with potassium, which had been isolated by the English chemist Humphry Davy the previous year. The reaction released boron from the boric oxide. The two Frenchmen beat the discoverer of potassium to the punch—Davy did not isolate boron until nine days later.

Boron is a lustrous, black material called a metalloid or semimetal. Boron is quite hard, but it is also brittle, so it cannot be used alone. It is sometimes added to steel and glass to strengthen those materials. Because it conducts electricity, boron is also used in the semiconductor (electronics) industry. One of its most common compounds is boric acid, $H_3BO_3$, which is a white solid. Boric acid is often dissolved in water and used as an antiseptic and eye wash.

Rocket scientists sometimes add boron to the solid propellants they hope will replace the heavier and bulkier liquid fuel and oxidizer used today. The liquids must remain frozen while

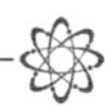

stored, while solid propellants can be stored under any conditions for any length of time. Boron particles, when burned, give off a great deal of energy.

Boron is found in Earth's crust in very tiny amounts, only 0.001 percent by weight. Rock contains 15,000 times more iron than boron, but boron becomes part of soil in sufficient amounts for most plants. Natural boron exists in two different isotopes—80.2 percent is boron-11, and 19.8 percent is boron-10.

The mineral tourmaline is one of the main sources of boron. Once tourmaline begins to break down, it often combines with organic (once-living) matter to form large molecular ions called oxyanions, especially $H_2BO^{3-}$. This is the form in which plants use the element.

## Boron in Living Things

Boron is an essential element in plants. This has been known since 1910, though it was not proved conclusively until 1923. For a long time there was no evidence that boron was essential to animals. It was not until 1987 that nutritionists proved that boron was an essential trace element for animals. It

**A scientist checks the root growth of kenaf plants, which are related to canola, to see whether they thrive in soil that contains considerable boron.**

appears to play a role in the body's metabolism, or use, of such macronutrients as calcium and phosphorus, but its role has not yet been fully determined.

Bones contain more boron than the softer organs of the body. The element may be necessary for calcium to do its work in promoting good bone growth. Like humans, chickens can suffer from the condition called rickets—a deformation of the bones—if there is not enough vitamin D (the "sunshine" vitamin) in the diet. By adding boron to chicken feed, rickets can be prevented in chickens.

Boron has been studied a great deal, but its exact role in human growth and development has not yet been determined. In the intestine, boron in food becomes $B(OH)_3$ with the addition of water. In this form, it moves through the blood.

Older women can develop a number of serious bone conditions. In these cases, they are sometimes given boron in the form of sodium borate, $Na_2B_4O_7$. It prevents calcium and magnesium (the main ingredients in bone) from being drained away.

Among the best sources of boron are avocados, peanuts, chocolate, wine, and dried fruits. Coffee, milk, and potatoes also add boron to the diet. There are no set requirements for the amount of boron a person should get every day.

In plants, boron regulates the use of hormones, which are messenger substances. Auxins are important plant hormones that instruct a plant to grow lengthwise, especially in response to light. Auxins are produced in young plant parts, such as flower buds. The chemicals travel downward through the plant to growth areas near the roots. Boron controls the movement of chemicals through cell membranes, a trait called permeability.

The tips of leaves on plants without enough boron die, while stem and root growth may be stunted. Leaves become curled and wilted. Plants without sufficient boron do not take up enough calcium from the soil.

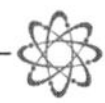

## Fluorine for Teeth and Bones

Fluorine, symbol F, element #9, is naturally a yellow-green gas. It is similar to chlorine, which follows fluorine in Group 17 (VIIA). Fluorine exists in Earth's crust at more than 0.05 percent.

When Humphry Davy was busy discovering elements in the early 19th century, he realized that an element with the properties of fluorine must exist. Researchers tried to isolate it, but fluorine is so reactive that the instant it was made, it formed a gaseous compound. Some of these compounds were poisonous and could eat away other substances. Fluorine was finally isolated in 1886 by Frédéric Henri Moissan, a French chemist.

Many compounds of fluorine are dangerous. Hydrogen fluoride gas (HF) and the acid made from it, for example, can burn breathing passages and skin. However, many fluorine compounds are useful because they resist other chemicals.

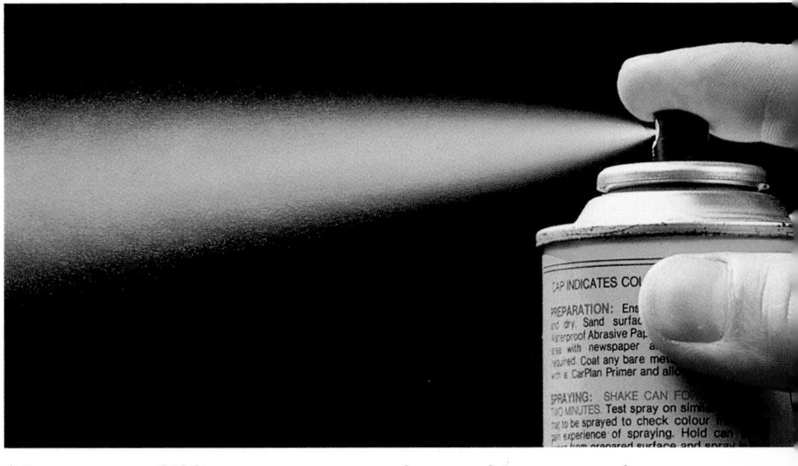


**Years ago, CFCs were commonly used in aerosol spray cans to carry other chemicals out.**

## Industrial Fluorine

Fluorine is among the chemicals that have caused environmental problems in our atmosphere. Years ago, chemists produced several different chemicals called chlorofluorocarbons, or CFCs. They were widely used in air-conditioning and freezing systems. They were also put in spray cans, such as those for paint. The gases carried the

paint out of the can but did not react with the paint.

But trouble was ahead. Released from a can or an air conditioner, CFCs rose in the atmosphere until they reached the stratosphere, in the region called the ozone layer. In the ozone layer, oxygen molecules consist of three atoms, the form called ozone, $O_3$. These molecules form a protective layer that prevents the sun's harmful ultraviolet rays from striking Earth. But those rays broke the CFC molecules apart, releasing chlorine atoms. Chlorine attacked the ozone layers, changing $O_3$ into $O_2$, which does not do the same protective job.

The fluorine part of the CFC molecules, however, does not play a role in this damage. Newer, less harmful versions of CFCs, called HCFCs, are still being produced in large quantities for the same uses.

Another chemical that contains fluorine is also known for not reacting with other chemicals. It is polytetrafluoroethylene, or PTFE, more commonly known by the trade name Teflon. It reacts with almost nothing and is so slippery that cooked foods and other substances slide across it.

The use of boron to propel rockets has an important drawback—boron oxidizes, forming a thin coating of boron oxide, which hinders ignition. Scientists have recently found, though, that fluorine will react with this oxide coating, letting ignition proceed rapidly. Fluorine has seven electrons in its valence shell, letting it quickly grab any electron available to complete and stabilize the valence shell.

## In the Teeth

In 1908, a dentist in Colorado Springs, Colorado, noticed that the children of his area seemed to have fewer cavities than most children. It wasn't until 1931 that the reason was discovered—the water in Colorado Springs has considerably more fluoride ions ($F^-$) than water in most places.

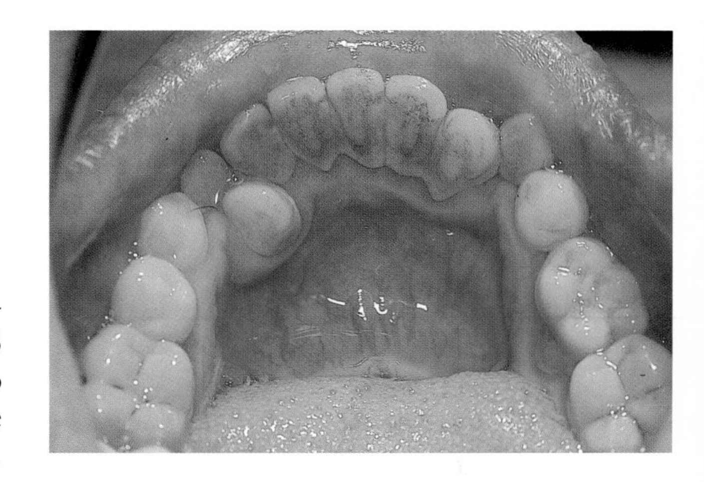

**Fluoride ions become part of the hard surface of the teeth, slowing attack by cavity-causing bacteria.**

Building on this knowledge, scientists began in 1945 to deliberately add fluoride to public water supplies in the form of sodium fluoride, NaF. About one part per million (ppm) fluoride in water helps teeth resist decay. In some places, this amount occurs naturally in water. The federal limit is 4 ppm, but few people get more than 2 ppm.

With too much fluoride, children's teeth may become mottled, or spotted. Tea plants are known to absorb fluoride from the soil, so someone who drinks a lot of tea may get too much fluoride. In individuals who receive massive exposure to fluorine, such as certain kinds of industrial workers, fluorine may take the place of calcium in bones. The bones become dangerously soft and crumbly.

Since fluoride has been added to public water supplies in the United States, the number of cavities in children's teeth has dropped more than 65 percent. About half of all children in the United States have never had a cavity.

Some people think that this benefit of fluoride is more cosmetic than essential—that it is a benefit to appearance more than to genuine good health. However, bad teeth can affect the way a person's body uses fuel. Infections in the teeth can harm the entire body. These facts make fluoride so important that it has been named an essential micronutrient.

# NUTRIENTS IN MICROGRAMS

Iodine is one of the few elements that sublimates, meaning that it goes directly from a solid to a gas.

At least five essential trace elements are needed only in tiny amounts called micrograms, which are millionths of a gram, or 0.0000000353 ounce! Even in such amazingly small amounts, these elements are important to us.

The effects of a deficiency of one of these elements—iodine—have been well known since ancient times. A shortage of iodine can cause a disfiguring bulge in the neck called a goiter. Another—selenium—is so difficult to locate that it was not even known to be essential until 1975.

## Iodine, the Seaweed Element

Iodine is an essential trace element for mammals but not for plants. However, it was originally discovered in plants. In 1811, a French chemist named Bernard Courtois was burning a mixture of seaweed, sod, and concentrated

sulfuric acid. After the mixture cooled, Courtois found that the vapor given off had solidified to violet-colored crystals. These crystals were identified as an element two years later by Joseph-Louis Gay-Lussac. At the same time, Humphry Davy named the new element from the Greek word *ioeides,* which means "violet-colored."

In ancient times, people ate seaweed to treat goiter, although they did not know why this worked. The reason this treatment worked was that seaweed contains iodine.

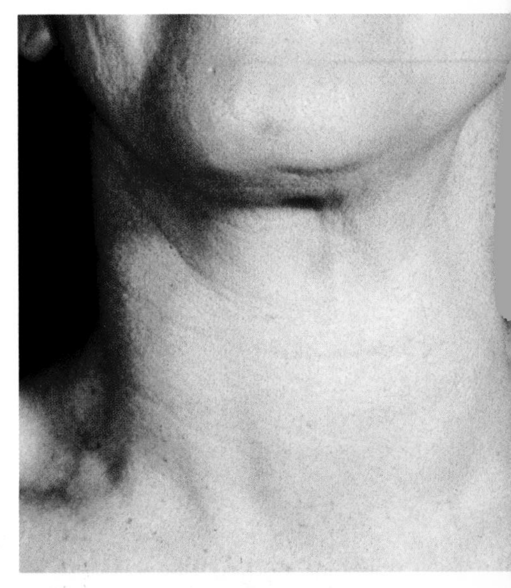

**Goiter is an enlargement of the thyroid gland, which shows at the front of the neck. It can be prevented by including iodine in the diet.**

Goiter is the enlargement of the thyroid gland, which lies at the front of the neck on each side of the cartilage called the Adam's apple. The thyroid was the first endocrine gland to be discovered. Endocrine glands are organs that release substances called hormones directly into the bloodstream

The thyroid gland is responsible for regulating all of the metabolic, or energy-using, functions of the body, such as growth and reproduction. The two hormones produced by the thyroid gland contain a great deal of iodine. They spread out through the body, entering every cell. They play a role in the making of proteins.

Lack of enough iodine in the diet causes goiter, along with such symptoms as weight gain and overall fatigue. A more serious consequence in children who grow up without sufficient iodine is possible mental retardation. The thyroid hormones also keep hair, skin, and nails healthy.

Once iodine was isolated as an element, physicians were able to create a more specific way to prevent goiter than eating

seaweed. They gave people the chemical potassium iodide, KI. People who live in areas of continents far from the sea, such as the Midwest of the United States, were particularly prone to goiter until potassium iodide was added to table salt.

Although most of the iodine in the body is found in the thyroid, it also occurs in the glands in the stomach that secrete gastric juices. In addition, there is iodine in the saliva given off by the salivary glands in the mouth.

For many years, tincture of iodine—a combination of alcohol, water, and iodine—was used on cuts and abrasions to kill bacteria. This method is used less frequently today because iodine can burn the skin. Some iodine-releasing compounds, called iodophors, are still used as antiseptics in hospitals.

## Iodine Isotopes

Iodine, chemical symbol I, is a nonmetallic element with atomic number 53. It exists at room temperature as diatomic, or two-atomed, molecules, written $I_2$. These molecules make a dark solid that sublimes—it evaporates directly to a violet-colored gas without becoming a liquid first.

Normal, stable iodine is iodine-127. At least two radioactive iodine isotopes—I-123 and I-131—are often used to trace small amounts of substances in

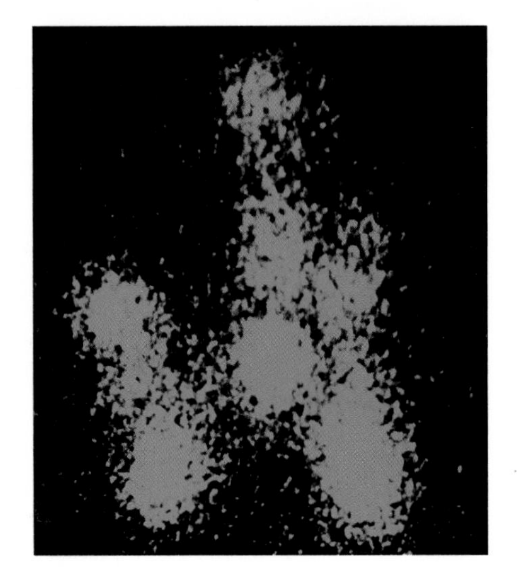

**I-131 is dangerous in the atmosphere, but tiny amounts can be injected into the body, where they accumulate in the thyroid gland, revealing it to X rays. This X ray shows a normal thyroid gland.**

the body. But I-131 has a much more harmful side to it.

Iodine-131 is an isotope produced by nuclear reactions. It continually changes, emitting harmful particles that can cause cancer. It is dangerous because it goes immediately to the thyroid gland. There it prevents regular iodine-127 from being taken up. The result may be cancer of the thyroid. When testing of nuclear weapons released radiation into the atmosphere, people all over the world were exposed to I-131. The rate of thyroid cancer rose. The only way to treat this cancer is to remove the thyroid gland. The patient then has to take medicines containing iodine for the rest of his or her life.

After atmospheric testing was discontinued, the largest known exposure to I-131 occurred when the Soviet nuclear power plant at Chernobyl in Ukraine exploded in 1986. I-131 was carried into the atmosphere and then settled on pastureland throughout Europe and Asia. It was carried even farther around the world by the wind. Fortunately, I-131 decays rapidly, becoming harmless within only a few days.

The people who had to drink the milk of cows who ate grass in the radioactive area were given potassium iodide as a protective measure. Even so, many cases of thyroid cancer occurred, especially in children.

## Chromium

French chemist Louis-Nicholas Vauquelin discovered chromium, element #24, in 1797. He was analyzing a compound called red lead, which was used as a red pigment, or coloring. He found that it contained an unknown metallic element. He named it after *chroma*, the Greek word for "color." It has the chemical symbol Cr.

Many of chromium's compounds come in various colors and so are used in various pigments. Oddly, it's the chromium in one mineral that makes the ruby gemstone red, while it's the chromium in a different mineral that makes the emerald gemstone green. The first laser, invented in 1960, used the chromium in a ruby to concentrate light rays.

Natural chromium is a mixture of four different isotopes—Cr-50, 52, 53, and 54. It is very common in the rocks of our planet. Chromium oxide, $Cr_2O_3$, is the ninth most common substance in Earth's crust.

Chromium behaves differently in different compounds. In some compounds, it gives up three electrons. When this happens, the chromium is called trivalent, written chromium(III). Chromium(III) is the kind of chromium found in organic compounds. Called serum chromium, it works outside the cells.

When Cr(III) is heated during industrial processes, it changes to chromium(VI), or hexavalent chromium. The compounds of Cr(VI) are generally toxic and harmful to the environment. Structurally, Cr(VI) is similar to phosphorus. Like phosphorus, it can enter the cells of the body with no trouble. Once in the cells, it can damage DNA.

## The Metal

Chromium is a bluish-gray metal that does not exist in nature. It must be removed from its ores, especially chromite, $FeCr_2O_4$. The $FeCr_2$ portion—called ferrochrome—is broken away by heating the mineral in an electric furnace. Then, without separating the iron and the chromium, ferrochrome is added directly to molten iron to make stainless steel. About 18 percent of stainless steel is chromium. It makes the iron stronger and more resistant to corrosion.

Sometimes, though, the iron is removed from ferrochrome to obtain pure chromium metal. Metallic chromium is used most often to plate, or coat, iron and steel surfaces that must not be allowed to rust, or corrode. The objects to be covered are dipped in the fluid of an electrolysis device, where the objects

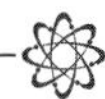

**These bits of pure chrome metal retain their shine when used to electroplate motorcycle parts.**

serve as electrodes that attract the chromium atoms. The shiny trim on cars and motorcycles is usually made by adding a chrome plating to another metal.

The toxic effects of chromium were first noticed in factories where the process of such chrome plating was carried out. Workers in plating plants and in chromium production plants were found to have much higher incidence of cancers of the lungs and nose passages than other people.

Chromium was originally regarded as a toxic substance that could be harmful. Then it was found that brewer's yeast (a primitive plant that makes beer ferment) was helpful to diabetics, people whose bodies had trouble using sugar. The part of brewer's yeast that was so helpful turned out to be chromium. Chromium doses can improve glucose tolerance, or the ability of the body to remove sugar from the blood and use it in cells for energy.

Some people think that taking a supplement of chromium can help them lose weight without harming their muscle mass. However, there is little scientific evidence for this.

People who do not eat enough vegetables may be slightly deficient in chromium, but regular servings of peas and corn will help. The FDA recommends that a person get 120 micrograms of chromium each day, though as much as 2 milligrams is regarded as safe.

**This environmental chemist is developing a way to safely use chrome-filled shavings from tanning leather to protect the environment.**

## Chromium in the Environment

Chromium(III) sulfate is used to turn cowhide into usable leather, a process that creates much hazardous waste. Because hides tend to be of uneven thickness, the bottom layer is shaved off and discarded. This discarded part contains considerable chromium. Scientists at the Agricultural Research Service have found that they can break down the chrome shavings with an enzyme called protease. The result is a high-grade protein gelatin that can be used for many things such as industrial packaging. To make the process even more beneficial, the chrome can be retrieved and reused.

Such recycling processes are helpful to the environment. More than half of all Superfund sites (areas where the federal government has found massive amounts of hazardous chemicals that need to be cleaned up) contain hexavalent chromium in their mix of chemicals. This kind of chromium easily enters water supplies, though trivalent chromium does not.

Erin Brockovich, the real-life version of the title character in the movie *Erin Brockovich,* was a single mother who worked as a legal assistant in a California law firm. She found evidence that a plant owned by the Pacific Gas and Electric Company was polluting the water supplies of the town of Hinkley with hexavalent chromium. Eventually, the company paid several hundred million dollars to the people of Hinkley for the damage to health allegedly caused by dumping the waste chemical.

Actually, chromium(VI) appears to be present in many private water supplies around Los Angeles. In 2001, tests showed levels of chromium(VI) more than 65 times higher than is considered safe. If allowed to accumulate in the body, chromium(VI) can harm the kidneys and may cause cancer. However, Cr(VI) in generally changed into Cr(III) by organic substances in the environment. Thus, harm to individuals usually occurs only near factories where chromium(VI) is being used.

## Cobalt

An American researcher, W. B. Castle, discovered in the 1930s that people who suffered from a disease called pernicious anemia—in which the body does not make enough good red blood cells—were missing some factor that was present in cow's liver. Their condition improved when they ate liver. This factor was called "antipernicious anemia" factor. It was later identified as vitamin B-12, or cobalamin.

The element cobalt plays a very small but vital role in vitamin B-12. It is a very complex molecule with a single cobalt ion, $Co^{+}$, at its center. Because all animals need cobalamin, cobalt has been named an essential micronutrient.

Without enough cobalamin, a person's cells do not produce DNA properly. As a consequence, sufficient red blood cells do not develop. The blood then cannot carry enough oxygen to all the cells of the body. The person experiences severe fatigue and can even suffer permanent nerve damage before the cause is discovered.

Cobalamin is produced naturally by microorganisms in the first stomach of cud-chewing animals. It is stored in the liver, which is why eating beef liver is so helpful to humans with pernicious anemia. Cobalamin is found mostly in meat, so vegetarians must make sure they obtain the nutrient from sources such as eggs and dairy products.

**Samples of radioactive cobalt are stored under deep water to keep the radiation from poisoning the surroundings. The samples give off an eerie blue glow.**

## The Blue Element and Its Isotopes

The name "cobalt" comes from the legendary goblin that German miners called the *kobold*. The miners were digging out a mineral that gave a brilliant blue color to glass. They named the mineral after the kobold.

The cause of the blue color was isolated by a Swedish chemist named Georg Brandt in 1735, but he did not know that it was an element. That discovery waited 45 years for the work of another Swedish scientist, T. O. Bergman.

Cobalt—symbol Co—is a metal, with atomic number 27. The metal is used in alloys for use in jet engines and various other industrial products, such as artificial joints and computers. Because of its importance in many military devices, cobalt is regarded by many countries as a strategic metal.

## Preserving Food with Cobalt

One of the most controversial uses of cobalt is the irradiation of food with the rays from a radioactive isotope, cobalt-60. The rays given off by this isotope have the ability to kill bacteria, mold, and small insects without harming the food that might house these pests. Irradiated fruits and vegetables can be stored longer than untreated foods.

The irradiation of food scares many people because it sounds "nuclear." However, the primary chemical that was used for preservation of fresh foods for many decades—methyl bromide, $CH_3Br$—was banned in the United States as of 2001. The bromine (Br, element #35) in the chemical acts like chlorine when it enters the upper atmosphere and damages the protective ozone layer.

Irradiation by cobalt-60 or another element, cesium (Cs, element #55), seems to be the best alternative. The rays pass through the food, and the sources of the radiation never touch it. The U.S. Food and Drug Administration approved irradiation of flour and other wheat products in the 1960s. The irradiation of meats was approved in 1992. Although some people object, such food-preservation methods will likely become even more common in the future.

## Molybdenum

Molybdenum has atomic number 42 and the symbol Mo. It was identified as an element by the Swedish chemist Carl Wilhelm Scheele in 1778 from the mineral known as molybdenite. Four years later, it was isolated by Peter Jacob Hjelm, Scheele's friend. Hjelm gave the new element its name, based on the Greek word *molybdos*, which means "lead." Scheele and others sometimes confused compounds of metallic molybdenum and the metal more properly called lead (Pb, for *plumbum,* element #82).

Molybdenum is a hard, silvery white metal. It is often used as an alloy with other metals because it contributes the ability to withstand high temperatures without melting. It also fights corrosion of the other metals. The metal adds strength to steel.

Very similar to silicon, molybdenum is often used in places where silicon might become too hot. Because molybdenum safely removes heat from the silicon, it is important in many electronic devices. Lighter in weight than many other metals, molybdenum is often used in aviation and space equipment.

One of the primary characteristics of molybdenum is its tendency to react with oxygen in the air to form molybdenum dioxide, $MoO_2$. At only 800°C (1,472°F), molybdenum dioxide sublimes, changing directly from a solid to a gas. Most often, this problem is solved by adding a coating of another material, usually one that is silicon based.

Metallic molybdenum is extracted from its main ore, molybdenum disulfide, $MoS_2$, called molybdenite. In its natural state, the element is a mixture of seven different isotopes, varying from Mo-92 to Mo-100. None of these isotopes exists in a large amount. Calculated together they give molybdenum an atomic weight of 95.94.

**Nodules of nitrogen-fixing bacteria form on the roots of some plants. The bacteria utilize molybdenum in the soil.**

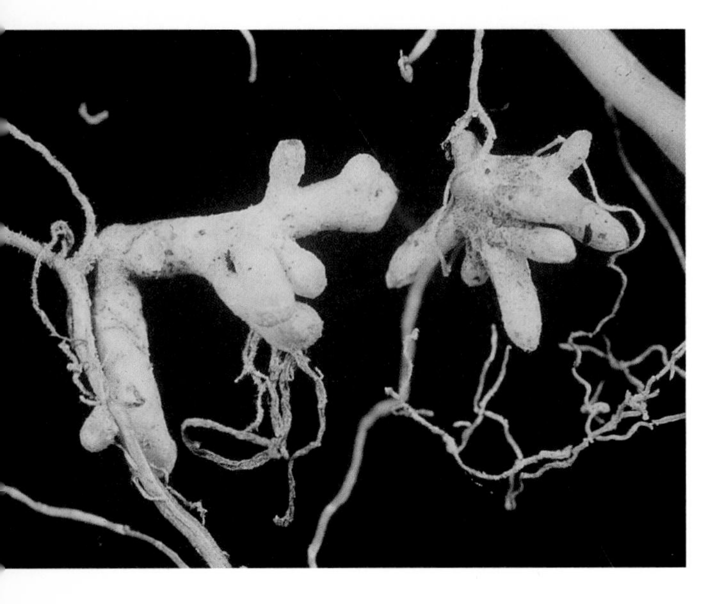

## Living Molybdenum

Plants depend on nitrogen to build proteins, but nitrogen has to change from the diatomic (two-atom) nitrogen molecule found in the atmosphere into an ion before a plant can use it. Most plants cannot make this change, called nitrogen fixing, happen. They have to depend on soil bacteria called *Rhizobia* to make the change. *Rhizobia* bacteria use molybdenum in soil in the process of fixing nitrogen. The only plants that can fix their own nitrogen are legumes, such as soybeans.

The molybdenum used by living things is the ion called molybdate, $MoO_4^{2-}$. It is required in the enzymes of many different plant and animal species. Three of these essential enzymes have been identified in humans. Each of these enzymes is necessary to metabolize certain amino acids.

Humans require a minimum of 25 micrograms of molybdenum a day. Whole grains and sunflower and pumpkin seeds are excellent sources of molybdenum, though most people generally get enough molybdenum in their normal diet.

Agricultural scientists know that molybdenum is important to certain crops, but they are not always certain of the amount needed. Lentils, which are legumes grown worldwide as a basic protein crop, require considerable molybdenum in the soil.

## Selenium

Selenium, chemical symbol Se and atomic number 34, is a metalloid, which means that it has some characteristics of metals and some of nonmetals. The photo of gray metal-like "pebbles" on the cover of this book shows only one of selenium's forms, or allotropes. It can also exist as a red powder or red crystal and as a black glassy material. These other three forms gradually become the gray metallic form when allowed to sit at room temperature.

The element was discovered by Swedish chemist Jöns Jakob Berzelius in 1817. He was analyzing the sediment that accumulated in the bottom of sulfuric-acid manufacturing chambers. At first, Berzelius thought that one mysterious substance was tellurium (Te, element #52), which lies next to selenium in Group 16, or VIA, of the Periodic Table. Then he realized that what he was finding was just different enough to be another element. He named the new element after the Greek word for "moon," which is *selene*. Today, selenium is generally acquired from the sediment in tanks where copper is refined by electrolysis.

Selenium is an important compound in the black powder used in photocopying machines. The metallic isotopes of selenium react to light by generating an electric current. In a related use, selenium is added to glass used in architecture because it helps keep out sunlight.

## Selenium on the Rocks

**Cowboys called this pea plant locoweed because it drove cattle "loco," or crazy.**

There is considerable selenium in the rocks of Earth's crust, though it is not distributed evenly. In the United States, the level of selenium in soil is highest in the upper Midwest and the Southwest. It is lowest in the Pacific Northwest and the Southeast.

The cowboys of the Wild West knew that a certain plant seemed to drive their cattle crazy when they ate it. They called the plant "locoweed." A member of the pea family, the plant absorbs considerable selenium from the soil. The excessive selenium is toxic to the animals.

In the United States, more than half the states have soil that is too low in selenium. This shortage requires farmers to buy selenium nutrients to add to their livestock feed. However, some farmers are taking advantage of the ability of another plant, canola—which is more commonly grown for its oil—to absorb selenium from soil. These selenium-overloaded plants can then be harvested as hay and fed to livestock in areas where the soil is very low in selenium.

Excess selenium in water has been shown to be harmful to wildfowl. It harms reproduction of ducks, geese, and other water birds. The baby birds growing inside eggs are often deformed and do not hatch. The young that do hatch from these eggs often do not grow adequately to survive into adulthood.

## The Good News of Selenium

Too much selenium can be harmful, but in smaller amounts selenium is important. It helps the body's immune system function to prevent disease.

Selenium was identified as being an essential element for human beings in 1975. Researcher Yogesh Awasthi found that it is a part of an enzyme called glutathione peroxidase, which acts in red blood cells. Selenium also serves as an antioxidant, preventing the oxidation of many other important substances. Oxidation leads to aging and hardening of tissues, which can be significant in heart disease.

The amount of selenium the average person gets in food is sufficient for the body's needs. A safe level of intake is up to 200 micrograms a day. It is found mainly in meat, fish, and grain products. Though selenium is lethal in large amounts, it seems to protect us from other, even more lethal metals, such as lead, mercury (Hg, for *hydrargyrum*, element #80), and arsenic (As, #33).

Recently, people have become interested in the role selenium might play in preventing various cancers. Selenium exists in foods in different forms. Selenate, $SeO_4^{2-}$, and selenite, $SeO_3^{2-}$, are ions found in broccoli, for example. They appear to play a role in the prevention of colon and other cancers. The form found in grains—selenomethionine—apparently does not.

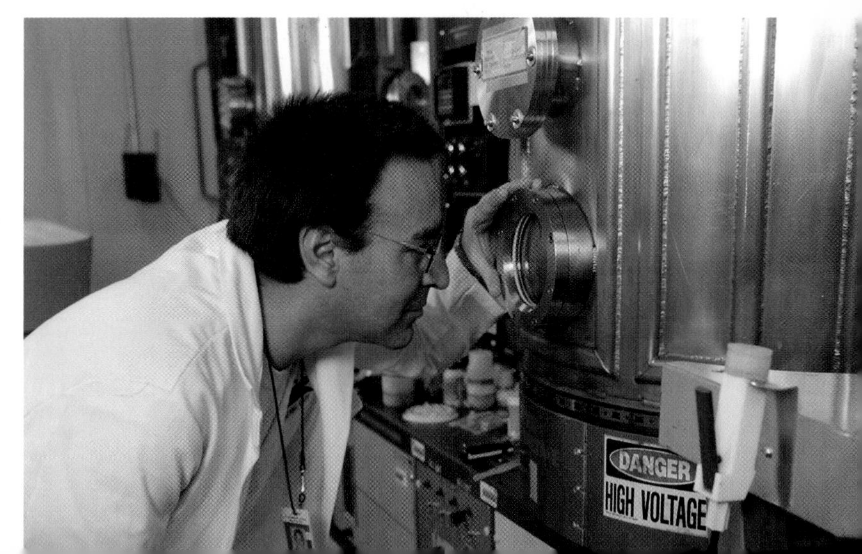


**The world's most efficient solar cell—used to produce electricity from sunlight—is being made from selenium, copper, and several other elements.**

# THE SEARCH GOES ON

For several hundred years, chemists have been discovering more elements essential to human life and health. The work is not done. Magnesium was not recognized as essential until 1950. It was the last of the macronutrients to be recognized. At that time, three of the elements now known to be essential were regarded as contaminants in food. They were selenium, chromium, and fluorine. Other elements are being studied, with scientists hoping to show that they are not only important but essential.

Many different research methods are used to study the foods we eat.

Trace elements are regarded as essential under these conditions:

1. If they are missing from a person's diet and then added in normal amounts, the individual's health improves.

2. If they are removed from the diet, health deteriorates.

3. If such deterioration is accompanied by a measurable change in the amount of mineral in the body.

4. If there is apparently no other mineral that could replace the normal

functioning of the one being considered.

5. If any qualified researcher who did the same investigation into a trace element would get approximately the same results.

Judith R. Turnlund of the Western Human Nutrition Research Center in Davis, California, researches trace elements. Among the things she wants to learn about each element are: " . . . where exactly does it go after we eat? How fast does it get there? How much do we store? How much do we lose?"

She often uses radioactive isotopes of the elements in her research. They can be followed in their paths from food to body to cells, to wherever the elements and their compounds go. She wants to know whether laboratory-created dietary supplements are as effective as natural elements and whether they can upset the body's natural mineral balance.

## Are They or Aren't They?

A number of elements being studied by various researchers may eventually prove to be essential but have not yet. For example, experiments have shown that an individual's health deteriorates if given a diet without nickel. But scientists have not determined yet what the actual function of nickel is or just how much is needed each day. They do know that an individual's growth may be retarded without enough nickel and that the body may fail to produce enough hemoglobin without adequate nickel. There is a fairly high concentration of nickel in the thyroid and adrenal glands. It may play a role in the metabolism of B-12 and an important substance called folic acid or folate.

Nickel can also cause problems. Nickel subsulfide, which exists in nickel refineries, can cause lung cancer. It is also found in tobacco smoke as nickel carbonyl. Less seriously, some people are allergic to nickel and develop a skin condition called nickel contact dermatitis. Often, people learn that they have this condition when they get their ears pierced with nickel studs.

**The absorption of nickel by plants is being studied in this greenhouse (left). A chunk of metallic silicon rests on a silicon wafer used in electronics (right).**

Silicon appears to play a role in bone formation. Most of the silicon in the body is in connective tissue such as bone, skin, and tendons. Chickens fed a silicon-free diet develop abnormalities in their bone structure. Beer, coffee, and water contain the highest amounts of silicon. In general, silicon comes from plant rather than animal sources, especially root vegetables such as carrots.

Vanadium, element #23, was discovered by a Mexican scientist, A. M. del Rio, in 1801. He originally called it *panchromium* because its salts came in many different colors (*pan* means "all" and *chrome* means "color"). He did not know that he had obtained an element. That discovery was made in 1830 by Swedish chemist Nils Gabriel Sefström.

Vanadium seems to play a role similar to insulin in the body's use of sugar. Doses of the element are often given to diabetics. Very little vanadium is absorbed by the body, so if it is essential, the requirement cannot be large. It may play a role in reproduction, since unborn rats do not develop when the mother's diet lacks vanadium. For humans, the best sources of vanadium are grains, along with mushrooms, shellfish, and various spices.

Vanadium exists in soil throughout the world, where it can cause problems for such plants as corn and soybeans, which do not need it. Because vanadium behaves chemically like phosphorus, these plants take up vanadium instead of the phosphorus they really need. Eventually, they come to suffer from a phosphorus deficiency.

## Are There Still More?

Scientists are finding that other elements may also prove to be essential. The identity of some of these, such as arsenic, is surprising. Generally thought of as a poison, arsenic may be needed for metabolism of methionine, one of the important amino acids used in building proteins. It can be found in dairy products, fish, cereal grains, grape juice, and spinach. Normal amounts of arsenic found in foods do not cause problems for most people. Arsenic is found in food in the form of organic arsenic compounds (those containing carbon). Only inorganic arsenic can be toxic, or poisonous.

The National Academy of Sciences plays an important role in determining what elements are considered to be essential. The U.S. Food and Drug Administration follows the NAS guidelines in making recommendations. In 2000, the NAS declared that there is clear evidence that arsenic, nickel, silicon, and vanadium play a "beneficial role" in some plant and animal species. However, they said there is not yet enough information to set Recommended Daily Allowances for these foods.

Even these elements may not complete the list of essential micronutrients. Barium (Ba, element #56), strontium (Sr, #38), and aluminum, for example, are also found within living tissues. Gallium (Ga, #31), cadmium (Cd, #48), lead, and tin are found in some lower animals. Some scientists believe that eventually they will find that such elements as antimony (Sb, #51), beryllium (Be, #4), germanium (Ge, #32), and tungsten (W, for *wolfram,* #74) will be found necessary in amounts so small that they would qualify as "micromicronutrients."

Are these elements essential to health? Scientists don't yet know. But someday, the "Nutritional Facts" panels printed on food packages may be much longer than they are now. Then we will know exactly what trace elements are in the food we eat.

# Iron and the Trace Elements in Brief

| Name | BORON | CHROMIUM | COBALT |
|---|---|---|---|
| Symbol | B | Cr | Co |
| Discovered or isolated by | Gay-Lussac and Thenard in 1808 | Louis-Nicholas Vauquelin in 1797 | Georg Brandt in 1737 |
| Atomic number | 5 | 24 | 27 |
| Atomic weight | 10.81 | 51.996 | 58.9332 |
| Electrons shells | 2,3 | 2,8,13,1 | 2,8,15,2 |
| Group | 13 (IIIA) | 6 (VIB) | 9 (VIIIB) |
| Usual characteristics | black crystal or red powder | brittle blue-white metal | hard, shiny, magnetic metal |
| Density (mass per unit volume) | 2.34 | 7.19 | 8.9 |
| Melting point (freezing point) | 2,076°C (3,769°F) | 1,900°C (3452°F) | 1,493°C (2,719°F) |
| Boiling point (liquefaction point) | 3,927°C (7,101°F) | 2,671°C (4,840°F) | 2,900°C (5,250°F) |
| Earth's crust (parts per million) | 10 ppm | 100 ppm | 20 ppm |
| Human body | 0.00007% | 0.000003% | 0.000002% |
| Stable isotopes | B-10 (19.8%), B-11 (80.2%) | Cr-52 (84%); Cr-53 (9.5%); rest is 50,54 | Co-59 |
| Radioactive isotopes | B-8,9,12,13 | Cr-51 | 16 known |

| COPPER | FLUORINE | IODINE | IRON |
|---|---|---|---|
| Cu (*cuprum*) | F | I | Fe (*ferrum*) |
| Known since ancient times | Frédéric Henri Moissan in 1886 | Bernard Courtois in 1811 & Gay-Lussac | Known since ancient times |
| 29 | 9 | 53 | 26 |
| 63.546 | 18.9984 | 126.903 | 55.847 |
| 2,8,18,1 | 2,7 | 2,8,18,18,7 | 2,8,14,2 |
| 11 (IB) | 17 (VIIA) | 17 (VIIA) | 8 (VIIIB) |
| reddish brown metal | yellowish gas or liquid | black solid non-metal | Soft gray malleable metal |
| 8.96 | 1.695 as gas | 4.93 | 7.87 |
| 1,083°C (1,981°F) | –18.8°C (–30.6°F) | 113.5°C (236°F) | 1,535°C (2,595°F) |
| 2,595°C (4,703°F) | –219.61°C (–363.3°F) | 184°C (363°F) | 3,000°C (5,432°F) |
| 60 ppm | 585 ppm | 45 ppm | 56,300 ppm; 4th most abundant |
| 0.0001% | 0.0037% | 0.00002% | 0.006% |
| Cu-63 (69%)<br>Cu-65 (31%) | F-19 (100%) | I-127 (100%) | Fe-56 (92%)<br>Fe-54, 57, 58 (8%) |
| 14 known | F-17, 18, 20-23 | 30 isotopes from I-110 to 140 | 12 known |

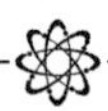

| MANGANESE | MOLYBDENUM | SELENIUM | ZINC |
|---|---|---|---|
| Mn | Mo | Se | Zn |
| Carl Scheele & Johan Gahn in 1774 | Carl Scheele in 1778; Hjelm in 1782 | Jöns Jakob Berzelius in 1817 | Known since ancient times |
| 25 | 42 | 34 | 30 |
| 54.938 | 95.94 | 78.96 | 65.39 |
| 2,8,13,2 | 2,8,18,13,1 | 2,8,18,6 | 2,8.18.2 |
| 7 (VIIB) | 8 (VIIIB) | 16 (VIA) | 12 (IIB) |
| hard, brittle silver metal | hard, silver-white metal | several: gray metal to red powder | bluish-white malleable metal |
| 7.44 | 10.2 | 4.5 | 7.14 |
| 1,245°C (2,273°F) | 2,610°C (4,730°F) | 221°C (430°F)] | 420°C (788°F) |
| 2,100°C (3,800°F) | 5,560°C (10,040°F) | 685°C (1,265°F) | 907°C (1,665°F) |
| 950 ppm | 1.2 ppm | 500 ppm | 7,100 ppm |
| 0.00002% | 0.00001% | 0.00007% | 0.0033% |
| Mn-55 (100%0 | 7, from Mo-92 to 100; Mo-98 is 24% | 6 between 74 and 82; half is Se-80 | Z-64 (48.6%), 66, 67, 68, 70 |
| 14 known, from Mn-49 to 62 | Mo-90,91,93,99 | 9 between Se-69 and Se-89 | 18 between Z-57 and 78 |

# Glossary

**acid:** definitions vary, but basically it is a corrosive substance that gives up a positive hydrogen ion (H+), equal to a proton when dissolved in water; indicates less than 7 on the pH scale because of its large number of hydrogen ions

**alkali:** a substance, such as an hydroxide or carbonate of an alkali metal, that when dissolved in water causes an increase in the hydroxide ion (OH-) concentration, forming a basic solution.

**anion:** an ion with a negative charge

**atom:** the smallest amount of an element that exhibits the properties of the element, consisting of protons, electrons, and (usually) neutrons

**base:** a substance that accepts a hydrogen ion (H+) when dissolved in water; indicates higher than 7 on the pH scale because of its small number of hydrogen ions

**boiling point:** the temperature at which a liquid at normal pressure evaporates into a gas, or a solid changes directly (sublimes) into a gas

**bond:** the attractive force linking atoms together in a molecule or crystal

**catalyst:** a substance that causes or speeds a chemical reaction without itself being consumed in the reaction

**cation:** an ion with a positive charge

**chemical reaction:** a transformation or change in a substance involving the electrons of the chemical elements making up the substance

**compound:** a substance formed by two or more chemical elements bound together by chemical means

**covalent bond:** a link between two atoms made by the atoms sharing electrons

**crystal:** a solid substance in which the atoms are arranged in three-dimensional patterns that create smooth outer surfaces, or faces

**decompose:** to break down a substance into its components

**density:** the amount of material in a given volume, or space; mass per unit volume; often stated as grams per cubic centimeter ($g/cm^3$)

**dissolve:** to spread evenly throughout the volume of another substance

**distillation:** the process in which a liquid is heated until it evaporates and the gas is collected and condensed back into a liquid in another container; often used to separate mixtures into their different components

**electrode:** a device such as a metal plate that conducts electrons into or out of a solution or battery

**electrolysis:** the decomposition of a substance by electricity

**electrolyte:** a substance that when dissolved in water or when liquefied conducts electricity

**element:** a substance that cannot be split chemically into simpler substances that maintain the same characteristics. Each of the 103 naturally occurring chemical elements is made up of atoms of the same kind.

**evaporate:** to change from a liquid to a gas

**gas:** a state of matter in which the atoms or molecules move freely, matching the shape and volume of the container holding it

**group:** a vertical column in the Periodic Table, with each element having similar physical and chemical characteristics; also called chemical family

**half-life:** the period of time required for half of a radioactive element to decay

**hormone:** any of various secretions of the endocrine glands that control different functions of the body, especially at the cellular level

**ion:** an atom or molecule that has acquired an electric charge by gaining or losing one or more electrons

**ionic bond:** a link between two atoms made by one atom taking one or more electrons from the other, giving the two atoms opposite electrical charges, which holds them together

**isotope:** an atom with a different number of neutrons in its nucleus from other atoms of the same element

**mass number:** the total of protons and neutrons in the nucleus of an atom

**melting point:** the temperature at which a solid becomes a liquid

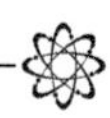

**metal:** a chemical element that conducts electricity, usually shines, or reflects light, is dense, and can be shaped. About three-quarters of the naturally occurring elements are metals.

**metalloid:** a chemical element that has some characteristics of a metal and some of a nonmetal; includes some elements in Groups 13 through 17 in the Periodic Table

**molecule:** the smallest amount of a substance that has the characteristics of the substance and usually consists of two or more atoms

**monomer:** a molecule that can be linked to many other identical molecules to make a polymer

**neutral:** 1) having neither acidic nor basic properties; 2) having no electrical charge

**neutron:** a subatomic particle within the nucleus of all atoms except hydrogen; has no electric charge

**nonmetal:** a chemical element that does not conduct electricity, is not dense, and is too brittle to be worked. Nonmetals easily form ions, and they include some elements in Groups 14 through 17 and all of Group 18 in the Periodic Table.

**nucleus:** 1) the central part of an atom, which has a positive electrical charge from its one or more protons; the nuclei of all atoms except hydrogen also include electrically neutral neutrons; 2) the central portion of most living cells, which controls the activities of the cells and contains the genetic material

**oxidation:** the loss of electrons during a chemical reaction; need not necessarily involve the element oxygen

**pH:** a measure of the acidity of a substance, on a scale of 0 to 14, with 7 being neutral. pH stands for "potential of hydrogen."

**pressure:** the force exerted by an object divided by the area over which the force is exerted. The air at sea level exerts a pressure, called atmospheric pressure, of 14.7 pounds per square inch (1013 millibars).

**protein:** a complex biological chemical made by the linking of many amino acids

**proton:** a subatomic particle within the nucleus of all atoms; has a positive electric charge

**radical:** an atom or molecule that contains an unpaired electron

**radioactive:** of an atom, spontaneously emitting high-energy particles

**reduction:** the gain of electrons, which occurs in conjunction with oxidation

**respiration:** the process of taking in oxygen and giving off carbon dioxide

**salt:** any compound that, with water, results from the neutralization of an acid by a base. In common usage, sodium chloride (table salt)

**shell:** a region surrounding the nucleus of an atom in which one or more electrons can occur. The inner shell can hold a maximum of two electrons; others may hold eight or more. If an atom's outer, or valence, shell does not hold its maximum number of electrons, the atom is subject to chemical reactions.

**solid:** a state of matter in which the shape of the collection of atoms or molecules does not depend on the container

**solution:** a mixture in which one substance is evenly distributed throughout another

**sublime:** to change directly from a solid to a gas without becoming a liquid first

**synthetic:** created in a laboratory instead of occurring naturally

**triple bond:** the sharing of three pairs of electrons between two atoms in a molecule

**ultraviolet:** electromagnetic radiation which has a wavelength shorter than visible light

**valence electron:** an electron located in the outer shell of an atom, available to participate in chemical reactions

**vitamin:** any of several organic substances, usually obtainable from a balanced diet, that the human body needs for specific physiological processes to take place

# For Further Information

## BOOKS


Atkins, P. W. *The Periodic Kingdom: A Journey into the Land of the Chemical Elements.* NY: Basic Books, 1995

Heiserman, David L. *Exploring Chemical Elements and Their Compounds,* Blue Ridge Summit, PA: Tab Books, 1992

Hoffman, Roald, and Vivian Torrence. *Chemistry Imagined: Reflections on Science.* Washington, DC: Smithsonian Institution Press, 1993

Newton, David E. *Chemical Elements.* Venture Books. Danbury, CT: Franklin Watts, 1994

Yount, Lisa. *Antoine Lavoisier: Founder of Modern Chemistry.* "Great Minds of Science" series. Springfield, NJ: Enslow Publishers, 1997


## CD-ROM


*Discover the Elements: The Interactive Periodic Table of the Chemical Elements,* Paradigm Interactive, Greensboro, NC, 1995


## INTERNET SITES

Note that useful sites on the Internet can change and even disappear. If the following site addresses do not work, use a search engine that you find useful, such as:
Yahoo:

http://www.yahoo.com

or AltaVista:

http://altavista.digital.com

A very thorough listing of the major characteristics, uses, and compounds of all the chemical elements can be found at a site called WebElements:

http://www.shef.ac.uk/~chem/we b-elements/

A Canadian site on the Nature of the Environment includes a large section on the elements in the various Earth systems:

http://www.cent.org/geo12/geo12/htm

Colored photos of various molecules, cells, and biological systems can be viewed at:

http://www.clarityconnect.com/webpages/-cramer/PictureIt/welcome.htm

Many subjects are covered on WWW Virtual Library. It also includes a useful collection of links to other sites:

http://www.earthsystems.org/Environment/shtml

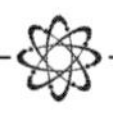

# Index

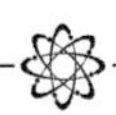